LEGACIES
SCIENCE AND TECHNOLOGY

Julian Rowe

Legacies

Architecture
Costume and Clothes
Language and Writing
Politics and Government
Science and Technology
Sports and Entertainment

Cover pictures: The Ferrari F355 (main picture) and one of its earliest ancestors, the Mesopotamian war chariot (inset picture). The legacy of the wheel continues to be at the forefront of modern technological development.

Series and book editor: Polly Goodman
Series designer: Liz Miller
Book designer: Malcolm Walker

First published in 1995 by
Wayland (Publishers) Limited
61 Western Road, Hove
East Sussex BN3 1JD, England

© Copyright 1995 Wayland (Publishers) Limited

British Library Cataloguing in Publication Data
   Rowe, Julian
   Science and Technology. – (Legacies Series)
   I. Title II. Series
   509

ISBN 0 7502 1309 4

Typeset by Kudos Editorial and Design Services, England
Printed and bound in Italy by G. Canale & C.S.p.A., Turin

Contents

Map of Ancient Civilizations	4
1. Science and Technology: Modern and Ancient	6
2. Ancient Middle East	10
3. Ancient Egypt	14
4. Ancient China	20
5. Ancient Greece	24
6. Ancient Rome	32
7. Ancient Americas	40
Science and Technology Time Line	44
Glossary	46
Books to Read	47
Index	48

Legacies are things that are handed down from an ancestor or predecessor. The modern world has inherited many different legacies from ancient civilizations. This book explores the legacies of science and technology from the ancient world.

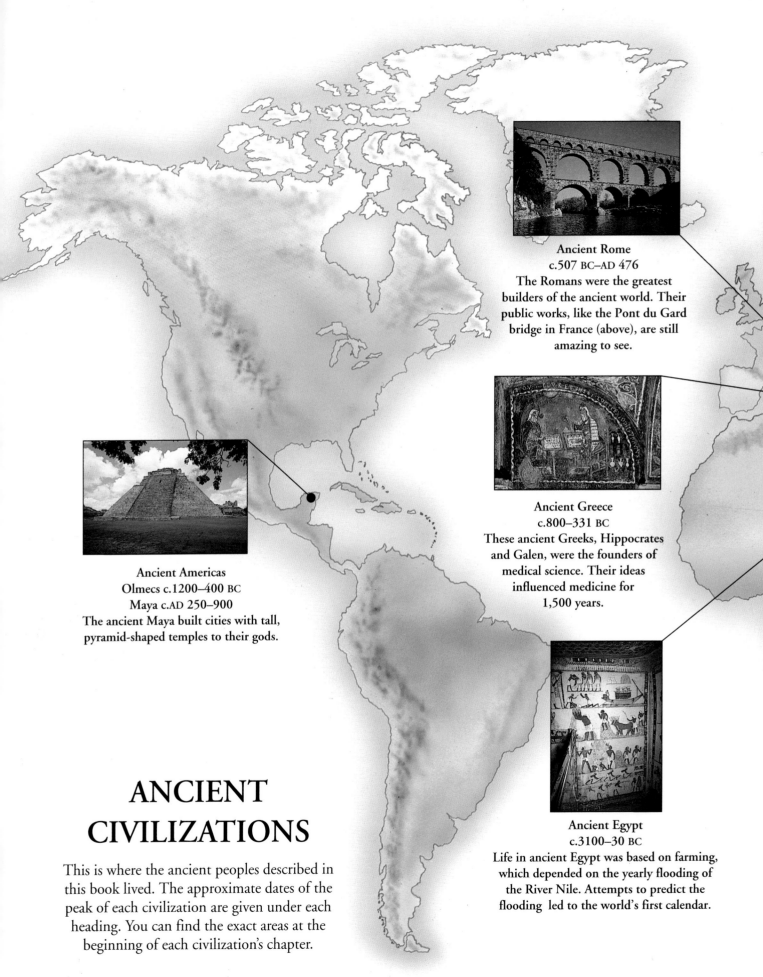

Ancient Rome
c.507 BC–AD 476
The Romans were the greatest builders of the ancient world. Their public works, like the Pont du Gard bridge in France (above), are still amazing to see.

Ancient Greece
c.800–331 BC
These ancient Greeks, Hippocrates and Galen, were the founders of medical science. Their ideas influenced medicine for 1,500 years.

Ancient Americas
Olmecs c.1200–400 BC
Maya c.AD 250–900
The ancient Maya built cities with tall, pyramid-shaped temples to their gods.

Ancient Egypt
c.3100–30 BC
Life in ancient Egypt was based on farming, which depended on the yearly flooding of the River Nile. Attempts to predict the flooding led to the world's first calendar.

ANCIENT CIVILIZATIONS

This is where the ancient peoples described in this book lived. The approximate dates of the peak of each civilization are given under each heading. You can find the exact areas at the beginning of each civilization's chapter.

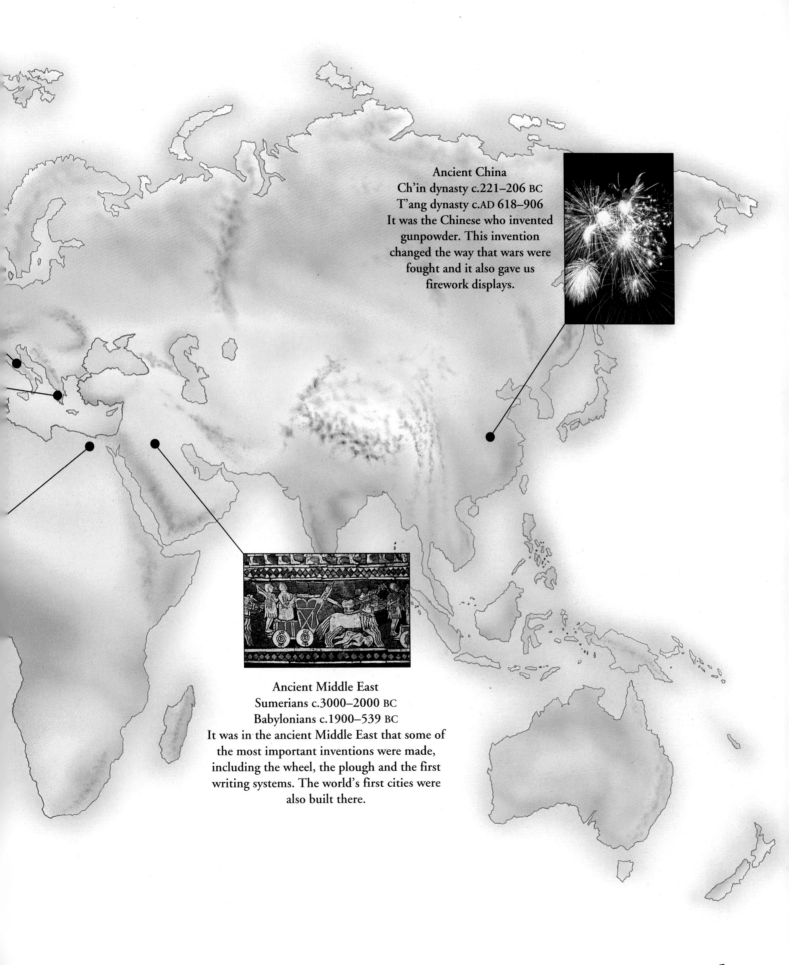

Modern and Ancient

chapter one

SCIENCE AND TECHNOLOGY: MODERN AND ANCIENT

Almost 12,000 years ago, as the great ice sheets that covered large areas of the earth melted, people used only stone tools and lived in caves or shelters. They used fire for cooking and to keep warm, and wore clothes made of animal hides. Where there were no caves to live in, people made large tents by stretching animal skins over frames of mammoth bones.

Today, many homes and buildings are centrally heated and meals can be cooked in seconds in a microwave oven. We wear lightweight clothes made from synthetic fibres. Although much of the science and technology we use today is very different from that of the past, many of the discoveries we use daily, such as heat and the fibres in clothing, can be traced back to our ancient ancestors and the civilizations they created.

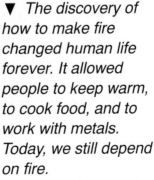

▼ *The discovery of how to make fire changed human life forever. It allowed people to keep warm, to cook food, and to work with metals. Today, we still depend on fire.*

So what are science and technology? The synthetic fibres that are used to make many of our clothes today were discovered by scientists working in laboratories. Scientists also discovered how to heat food with microwaves. Scientists study the natural world and try to explain what they see happening around them. Technology is how science works in practice. Important inventions or discoveries in science and technology change how we live.

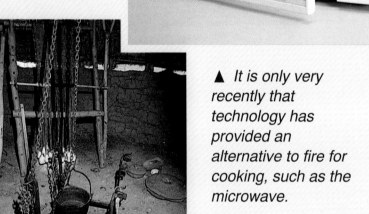

▲ It is only very recently that technology has provided an alternative to fire for cooking, such as the microwave.

◀ This is what the inside of a Celtic Iron-Age hut in Europe would have looked like almost 3,000 years ago. The fire was the centre of home life, providing heat for cooking and warmth.

For example, when metalworking (metallurgy) was discovered, it began the end of the Stone Age over most of Europe and the Middle East. People no longer depended on tools and weapons made of bone, antler, ivory, or stone. For a long time, they had found lumps of copper and gold and hammered them into shape with primitive stone tools. But most metals are locked up in rock-like ores. It took a scientific discovery – that a fierce, hot fire is needed to extract the metal – before useful objects of copper or iron could be made from ores. But to make the fire hot enough, bellows had to be invented. This invention is an example of technology.

The inventions that we use and improve on today are one type of legacy. But another type of legacy comes from the ideas our ancient ancestors had about the world around them. Nuclear physics and modern medicine, for example, grew from some of the first ideas about what things are made of and how the human body works.

Modern and Ancient

▲ An ancient Egyptian harvesting scene. The farmers walk their oxen over the wheat to break off the grains from the stalks. Then it is winnowed – tossed into the air so that the heavier grains fall to the ground while the stalks blow away.

FOOD AND FARMING

In most parts of the world today, getting food is simply a case of a trip to the local supermarket. There it is easy to buy enough food to feed a family for a week. Such shopping is only possible today because of modern farming methods and transport. The story of farming starts in about 8000 BC in Mesopotamia, in an area called the Fertile Crescent. This is in modern-day Iraq and Jordan.

Farming became possible because of the natural cross-breeding of the genes in two forms of wild wheat. Genes are the material in the bodies of plants and animals that carry and pass on characteristics. The cross-breeding of the two original wheats produced a different type of wheat called Emmer. Emmer wheat, the ancestor of the wheat we eat today, produced a much heavier crop than the wild varieties. This meant that people no longer had to forage for food across large distances because enough wheat to feed them could be grown in one place. Hunter-gatherers could settle in one place and become farmers.

Wheat and barley were the first plants to be grown and cultivated by humans. The first farmers improved their crops by planting the seed taken from their best plants – those with the heaviest crop. They also domesticated a variety of animals, such as goats, sheep, cattle and donkeys, just like the variety in farms today. Goats were a popular farm animal because they needed little food but provided milk. Sheep provided wool, and cattle and donkeys could be used as draught animals to transport materials.

When people first began to grow more food than they needed for themselves, others in the community could specialize in new activities other than finding food. In modern industrial countries, not many people are needed to work the land. A few farmers can produce enough food for everyone.

Genetics
Genetics is the study of heredity, characteristics inherited through genes. Today, the people who study genetics, called geneticists, try to find ways of improving plants and animals by altering their genes. This is called genetic engineering. Plants can be improved in this way to increase the crop they give.

▼ A modern combine harvester reaps the corn, threshes (breaks it up) and then winnows it. One person driving a combine harvester can do all the harvesting work of an ancient Egyptian village.

chapter two

ANCIENT MIDDLE EAST

The first known cities were built beside rivers in Mesopotamia by the ancient Sumerian civilization, about 3500 BC. There, in the valleys between the Tigris and Euphrates rivers, it was easy for people of cities such as Nippur, Uruk and Lagash to use the rivers for travel, for trade or for fishing. Date palms were plentiful and wildfowl teemed in the marshes. The rivers were also a source of fresh water. A third of the world's population live in cities today, many of which are built on ancient city sites, beside rivers, such as Rome in Italy and Istanbul in Turkey.

IRRIGATION

The inhabitants of cities have always needed regular supplies of food. Growing this food, in ancient as well as modern times, has needed technology. Plants need water to grow. For the Sumerian civilization of the ancient Middle East, the Tigris and Euphrates rivers flooded regularly.

Çatal Hüyük
The ancient city of Çatal Hüyük in modern-day Turkey, was built about 7,000 years ago. It had a population of at least 5,000 people. They lived in houses made of mud, brick and wood and were the first people to use irrigation. The wealth of the community came from the trade of obsidian, a volcanic rock used to make tools.

This left a layer of rich soil on the plain in between them, but water was still needed for the crops. So the Sumerians and their successors found ways to irrigate, or supply water to, their crops. They constructed earth dams, sluices and canals, which made it possible to lead water to a network of furrows which criss-crossed the fields. These furrows irrigated the crops directly. Another method of irrigation was to dig deep channels around plots of land, and keep them filled with water. This irrigation technology ensured a supply of food for the citizens of the world's first cities.

Today, dams are still used to store water for irrigation. The stored water is also a source of energy, because it can be used to drive electric generators in hydroelectric power stations. The Grand Coulee dam, near the head of the Columbia River in Washington State, USA, opened in 1942. It is the largest concrete dam in the world. The dam provides water for the Coulee irrigation system and produces 6.5 million kilowatts of electricity a year.

▲ An Assyrian carving, made almost 3,000 years ago, showing an early dam on a river. The man is working a lock – a device for controlling the flow of the water from the higher to the lower level.

▼ This modern dam is part of a hydroelectric power station in New Zealand.

Ancient Middle East

▲ This ancient Sumerian war chariot was one of the first vehicles to use wheels. These first wheels were heavy and solid, which would have made them move roughly along the ground.

THE WHEEL

The wheel is another legacy we owe to the ancient Middle East. Some of the first people to use the wheel were potters in Mesopotamia about 5,000 years ago. They used the wheel to spin clay, which they shaped into pots. The first cart-like vehicles also appeared in Mesopotamia but much later, in about 3200 BC. With heavy, solid wheels, these vehicles were used by the Sumerian army as primitive chariots. Wheels with spokes came even later. They were lighter than the solid wheels and first appeared around 2000 BC. These spoked wheels were used as farming vehicles and war chariots. Then, around 100 BC, wagon-makers from Denmark, in Europe, experimented with the wheel by using roller bearings. Roller bearings made the wheel turn more smoothly by reducing the friction in the middle of the wheel as it turned.

The legacy of these early wheel-builders surrounds us today in many different forms, from industry to sport. The international motor industry produces millions of vehicles each year, and the latest sports, such as Rollerblading, find new ways of using the wheel. From privately-owned cars and motorcycles to giant jet airliners, all kinds of transport depend on the wheel.

SEVEN-DAY WEEK

Astronomers in the ancient city of Babylon (modern-day Iraq) produced a calendar for keeping track of time. It is from the Babylonians that we have inherited a seven-day week, with the days named after the planets. Seven was a sacred number in ancient Babylon, perhaps because the Babylonians only knew of six planets and the sun. So we have Moon day, Mars day, Mercury day, Jupiter day, Venus day, Saturn day and Sun day. Our hours and minutes are another legacy from ancient Babylon. Mathematicians at the time used sixty as a counting unit, so we have sixty seconds in a minute and sixty minutes in an hour, not to mention 360 degrees in a circle as a result!

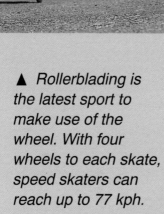

▲ *Rollerblading is the latest sport to make use of the wheel. With four wheels to each skate, speed skaters can reach up to 77 kph.*

▼ *The Ferrari F355 – a high-performance sports car that can go from 0–100 kph in 4.7 seconds.*

Ancient Egypt

chapter three

ANCIENT EGYPT

The plough is a piece of farming technology, used all over the world today, which can be traced back to ancient Egypt. Ploughs are used to till, or break up the soil after a crop has been harvested. This forms a new seed bed ready for new seeds to be planted. A modern plough has two blades: a cutting blade and a V-shaped blade. The cutting blade bites into the soil first, then the V-shaped blade turns the soil over.

▲ *The blades, or ploughshares, of a modern plough turn the soil in a field of maize.*

Ancient Egypt

The first ploughs were little more than hand tools which scratched the surface of the soil. Then, oxen were used to pull a type of plough called a scratch plough. A model of an ancient Egyptian scratch plough, worked by a farmer and oxen, was found in an Egyptian tomb that dates back about 4,000 years. A similar type of scratch plough can still be seen in action on the banks of the River Nile in Egypt.

Improved ploughs with wheels were invented in the sixth century in Europe. They could turn over heavy soil and needed a team of four, six or eight oxen to pull them. Today's high-powered tractor pulls a plough that tills six furrows at a time. The driver uses a computer to control the plough and the amount of fuel consumed.

▲ Many Egyptian farmers still use oxen and a type of scratch plough, just as they have been doing for thousands of years.

◀ This ancient piece of papyrus shows oxen pulling a scratch plough across a field in the underworld. This was the place where Egyptians believed they went after death. The ancient Egyptians could not imagine a life without farming, even in the next world.

Ancient Egypt

THE BIGGEST TOMBS

Accurate construction skills are another technology legacy of the ancient Egyptians, who used these skills to build great pyramids as tombs for their pharaohs. The pyramids, such as the Pyramids of Giza, were built with such skill that they impressed Greek and Roman tourists 2,000 years ago, just as they impress visitors today.

The largest of the pyramids, the Great Pyramid of Cheops, measures 147 m high and has an area of 250 m square. It was made from over 2 million limestone blocks, each weighing 2.5 tonnes on average. These mighty blocks were dragged into position over log rollers by gangs of workers.

The limestone and granite blocks often had to be cut in quarries far from the pyramid sites. They were then transported with great skill on barges up the River Nile. The blocks

▼ This is an artist's idea of how the pyramids must have been constructed. The pharaoh is being shown the plans by his architect on the right. Just below them are a set square and plumbline, tools that were used by the architect to plan the pyramid. Beside the tools, the workers' supervisor cracks a whip.

were then finished at the building site using hammers and chisels made of dolerite, a very hard stone, and saws and drills made of bronze. A modern tool kit still contains items that are similar to these ancient tools.

The pyramids were very precisely planned and accurately constructed. The pharoah's architect would have designed the pyramid and planned the square base. To make measurements, the ancient Egyptians used a measuring cord – a long cord made from palm fibre, and knotted at regular intervals. The architect would probably have used this piece of basic technology when planning the pyramid base. The knots on the measuring cord separated units of length called cubits, which were various lengths of the forearm. The base of the Great Pyramid of Cheops has four sides measuring 440 cubits each. Each cubit equals about 52 cm. Today, architects and civil engineers use high-tech computers to draw up building plans, but the pyramids of Egypt and the skill of the people who planned and built them are still wonders of the modern world.

▲ High-tech computers make very accurate plans and measurements for use in modern architecture and civil engineering.

◄ This ancient limestone model, from 1400 BC, shows an Egyptian surveyor holding a measuring cord, rolled up into a coil.

BEER

Today an enormous variety of beer is displayed in supermarkets and bars. In addition to beers brewed locally, beers from all over the world are available. In ancient Egypt, beer was just as popular and it was drunk by everyone. Much thicker than modern beer, ancient Egyptian beer was more like a drinkable bread, and it was also used as a medicine. Brewing beer started out in the same way as making bread, by mixing flour, water and yeast to make a dough. Then the dough was cut up and placed in water. This process produced a very strong beer which the ancient Egyptians flavoured with spices and bitter substances.

BREAD AND YEAST

Many ancient civilizations made bread of one sort or another, but it was the ancient Egyptians who first made dough using yeast, an ancestor of modern Western dough. The Egyptians discovered that yeast mixed with flour, water and air produced a sour dough that made a softer and lighter bread. The Egyptians perfected the art of baking and produced white bread from wheat. White bread was eaten by the rich. Some bread was shaped into long rolls and decorated with seeds, just like some modern loaves.

▶ The first stage of bread-making, an Egyptian woman grinds her wheat into flour, using a stone pushed backwards and forwards. This method of grinding is still used in parts of Africa.

◀ Large industrial ovens turn out hundreds of loaves of bread a day in modern bakeries. Bread has been a basic human food for thousands of years.

CALENDAR

A calendar is a way of marking the beginning and length of a year and dividing it into sections. Today's Western calendar of 365.25 days, which is in use today, can be traced back to the detailed calendar the Egyptians invented around 4300 BC.

At first the Egyptians and other ancient peoples used a lunar calendar – one that relied on the phases of the moon. However, lunar calendars were not fixed, and every three years or more there would be an extra month. This would have been very disruptive for the highly organized Egyptian civilization. So, by observing the movements of the stars at night throughout the year, and by noting the time of the annual flooding of the Nile, they worked out a fixed, more reliable calendar. The Egyptians were the first people to work out a calendar not based on the moon, and to understand the length of the solar year – 365.25 days – based on the observation of the star Sirius.

Changing time
Calendars used to be very inaccurate and they would gradually become out of step with the seasons. In 1752 in Britain, the calendar was adjusted by eleven days. This adjustment caused riots: people believed that their lives were being shortened.

Ancient China

chapter four

ANCIENT CHINA

COMPASS

Magnetic compasses are used as a guide for navigation today, to show the direction of north, south, east and west. They react to the earth's magnetic field because they are made of naturally magnetic material. The first magnetic compass was invented by the ancient Chinese, in the first century AD. The 'south-pointing spoon', as it was called, was used by fortune tellers to find the luckiest sites for buildings to be constructed. This was part of an ancient Chinese belief that the earth has natural lines within it – and that these lines should influence where buildings should be planned. The 'south-pointing spoon' consisted of a spoon-shaped pointer containing a piece of lodestone, a naturally magnetic rock. The spoon rotated on a highly polished plate when the lodestone reacted to the earth's magnetism. It was at least 1,000 years before the Chinese used the compass for navigation, and it was not until the twelfth century that the discovery was passed on to the West by Chinese sailors.

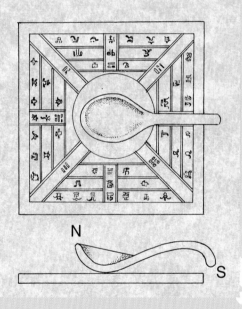

South-pointing spoon

▲ The 'south-pointing spoon' was the first magnetic compass. Magnetism, a powerful invisible force, seemed magical to the Chinese. The direction in which the spoon's handle pointed was used by fortune tellers to show lucky sites for buildings.

EARTHQUAKE!

Today, scientists record the features of earthquakes using a seismograph. The first piece of technology to record earthquakes was invented in ancient China in about AD 200. This device, called a seismoscope, was invented by

20

Ancient China

◄ Chang Heng's seismoscope has not survived, but this reconstruction of it shows us what it looked like and how it worked. The frog that the ball dropped into showed the direction of the earthquake.

an astronomer, geographer and mathematician named Chang Heng, and it recorded the direction of the earthquake. The seismoscope was made of bronze and looked like a wine jar, approximately 2 m in diameter. A heavy pendulum was suspended inside the bronze casing, which was free to move in any direction when an earthquake struck. The pendulum's movement caused a ball to fall from the jaws of a model dragon into the mouth of the nearest in a circle of model frogs below. This showed the direction the shock was coming from.

Modern seismographs work on exactly the same principle as their ancient ancestor, using a suspended pendulum. Instead of dragons and frogs, the modern pendulum controls a pen which draws a wavy line on a roll of moving paper. The pen trace of the seismograph shows exactly how strong an earthquake was.

▼ This is a modern seismograph with its roll of paper, which records both the direction and strength of an earthquake. The strength of earthquakes is measured on the Richter scale. The largest recorded earthquake measured nine on the Richter scale.

21

CALCULATIONS

Calculations that are too complicated to make in the head can be done very quickly today with a calculator. Electronic calculators depend on a microchip and are a recent invention. But fast methods of calculation were possible in ancient times too. The ancient Chinese abacus was a device that used counters which were strung on wires and placed in a frame. A bar across the frame separated it into a bottom section and a top section, with each bottom wire holding five counters, and each top wire holding two. Each of the five counters in the bottom section stood for one unit, and each of the two counters in the top section stood for five units. This device, which in the hands of an expert operator can be faster than an electronic calculator, is still used in China, Japan and Russia.

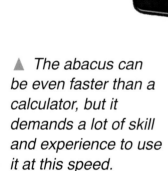

▲ The abacus can be even faster than a calculator, but it demands a lot of skill and experience to use it at this speed.

FIREWORKS

Fireworks are a popular way of celebrating special occasions all over the world today. They work by using a type of gunpowder made from substances such as potassium nitrate and finely ground charcoal. The ancient Chinese were the first to use fireworks, such as firecrackers, which they used to ward off evil spirits. They also invented gunpowder. The first fireworks did not use gunpowder. Firecrackers were made from hollow pieces of bamboo, blocked at both ends, which exploded when thrown into a fire. Gunpowder was invented later, by about AD 800. It was made from a mixture of charcoal, saltpetre and sulphur, very like the mixture used to make modern fireworks. As the gunpowder burned, it created smoke and gases which shot out of the firework and drove it forward.

▲ *Next time you see a firework display, think of the ancient Chinese inventors who first made gunpowder. They must have been very surprised when they saw what it did!*

Ancient Greece

chapter five

ANCIENT GREECE

▼ *This is a medieval painting of Galen and Hippocrates. Until the sixteenth century, doctors believed that these two Greeks had explained all there was to know about medicine.*

MEDICINE

Some doctors today are required to take the Hippocratic Oath, a code of conduct which requires medical professionals to treat all sick people fairly and not to tell confidential information about patients. This oath comes from Hippocrates, the most famous ancient Greek doctor, who was born on the island of Cos, about 400 BC.

Hippocrates is often called the 'father of medicine' because his ideas formed the basis of modern medical science. In ancient Greece, doctors' methods were often based on religion and superstition rather than on scientific knowledge. Hippocrates, however, believed that all medical theories should be tested and that medical knowledge should be based on practical observation. Modern medical teaching follows this belief: medical students learn about the human body today by cutting it up, observing it and making notes on what they find. Hippocrates wrote many books containing information on medical matters. He also wrote a code of principles for teachers of medicine and their students, which is where the Hippocratic Oath comes from.

GALEN AND BLOOD

We take it for granted that our blood circulates around our bodies, but only because we have been told that it does. Ancient doctors used to think that the arteries in the human body circulated air, not blood. The first doctor to show that it is blood, not air, that is contained in the body's arteries was the Greek physician Galen. He was also the first doctor to use the pulse as a test of health. Appointed chief physician to the gladiators in AD 157, Galen later became both friend and physician to Emperor Marcus Aurelius. He wrote eighty-three medical text books, in which he discussed nutrition and human digestion.

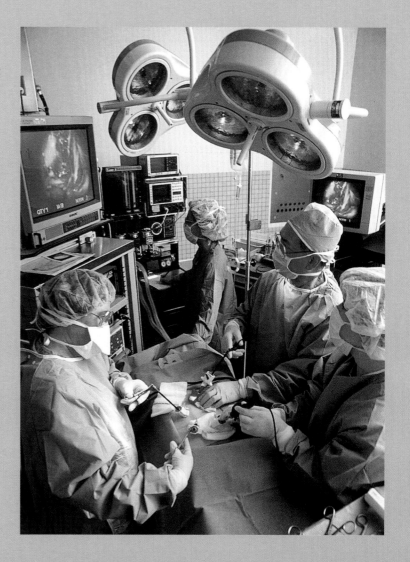

▲ Modern surgery is only possible because of the slow gathering of medical knowledge over time, a process that was started by Hippocrates and Galen.

Drugs from plants
Modern drugs come from many sources. Some are made in the laboratory, others come from micro-organisms, but many come from plants. Medical drugs from plants are a legacy from ancient Greece, where physicians used herbal drugs obtained from root-gatherers called 'rhizotomoi'. Today, new drugs are often discovered in the rainforests of the world, where there are many medically valuable plants. This is one important reason for rainforest conservation.

Ancient Greece

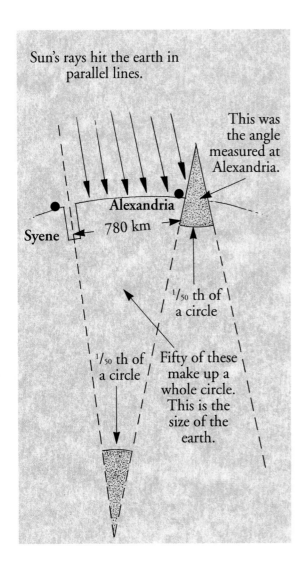

▲ Using a home-made sundial (a stick standing upright in a bowl), Eratosthenes measured the angle of the sun at Alexandria. He then used this angle, and the distance from Syene to Alexandria, to work out the size of the whole earth.

MEASURING THE EARTH

The first person to measure the size of the earth was a brilliant Greek mathematician called Eratosthenes, who lived in Alexandria, Egypt. In around 230 BC, Eratosthenes measured the distance around the earth using the angle of shadows cast by the sun. Eratosthenes found out that on 21 June each year, the midday sun was directly overhead in the town of Syene, to the south of Alexandria. He knew this because a traveller had told him that the sun had lit the bottom of a deep well there. Eratosthenes knew that at the same time, the sun was not overhead in Alexandria, his home town, because there were always shadows in Alexandria. So he waited until midday on the next 21 June and, using a home-made sundial, Eratosthenes measured the angle of the shadow cast by the sun at Alexandria. He found that it was one-fiftieth of a circle. This meant that the distance between Syene and Alexandria was one-fiftieth of the distance around the earth. Since Eratosthenes knew this distance, by multiplying it by fifty, he came up with the distance around the earth – about 39,000 km. This was amazingly accurate. The real distance is just over 40,000 km.

PLANETS AND STARS

Planetariums today project computer-controlled images of the stars and planets on to a curved screen for people to see models of the sky at night. One of the first models of the universe that could predict the positions of the stars was constructed by the ancient Greeks in about 87 BC. The model, called the antikythera instrument, was found on an ancient shipwreck near the Greek island of Antikythera. It was composed of at least twenty-five bronze gear-wheels, now very corroded, which interlocked and were set in motion by an axle that caused pointers to move on a set of dials. These indicated the position of the sun, the moon and some of the planets.

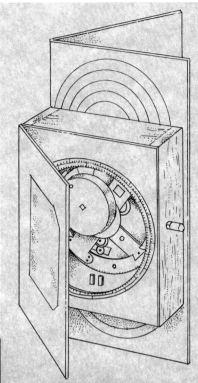

◀ **Antikythera Instrument**
This is a reconstructed drawing of the ancient antikythera instrument. The axle that turns the pointers sticks out on the right-hand side. Inside the wooden case, you can see the interlocking wheels, which show the movements of the heavenly bodies like a clock.

Pythagoras
Pythagoras, who was born on the Greek island of Samos in about 560 BC, was an important figure in ancient Greece. He investigated mathematical and musical theory. Pythagoras believed that the earth was fixed, and that the planets moved around it on seven spheres, one inside the other. He related the movements of these spheres to musical theory, and talked about the 'harmony of the spheres.'

▲ *The Greeks were skilled astronomers who plotted the movements of the planets. But they were held back by one huge error. Like all ancient peoples, they did not realize that the earth itself is a planet, moving through space.*

ARCHIMEDES, 'FATHER OF MECHANICS'

Archimedes was a mathematical genius in ancient Greece. Born about 287 BC, one of his greatest legacies is mechanics (the science of machines) and levers. Levers are devices that help lift heavy objects. They use a bar that pivots, or balances on, a fulcrum, or middle-point. Although all kinds of levers had been in use before Archimedes' time, he set out exactly how a lever could be used. He saw clearly that levers could be used to move extremely heavy loads without using much effort. He invented the Archimedes' screw, a machine for lifting water. Machines like this are still used for lifting water in many places.

Archimedes also discovered how to measure an object's volume and density. He discovered density almost by accident. Hiero, the King of Syracuse, suspected that his goldsmith had cheated by adding some silver to his new gold crown. So he sent for Archimedes and asked him to solve the problem. While Archimedes was getting into his bath, he noticed that the water level rose. He realized that the amount of water that an object displaces is the same as the volume of the object.

▲ Archimedes of Syracuse was born into a noble family and was educated in Alexandria, a city in what is now Egypt.

'Give me a place to stand and I will move the earth.' – Archimedes of Syracuse (c.287–212 BC)

Archimedes saw that he could use this discovery to measure the volume of the king's crown. He was so exicted that he ran through the streets naked, shouting *'heureka!'*, which means 'I have found it!'

The goldsmith had been given a lump of gold by Hiero to make his crown. Archimedes found an identical lump and measured its volume by sinking it into water. When he sank the crown into water, it displaced more water than the gold, therefore it had a larger volume. The goldsmith had taken some of the gold for himself and replaced it with silver. Because gold is denser than silver, he had to add a larger amount of silver than the gold he had stolen. This made sure that the weight of the crown was the same as the original lump of gold.

Totally immersed
Archimedes was killed at the siege of Syracuse by a Roman soldier, because he was so involved in a mathematical problem that he ignored the soldier's challenge. The Roman general Marcellus gave Archimedes an honourable burial to show his respect for Archimedes' genius.

◀ *Archimedes' inventions are still in use today. This pumping station in Holland uses an Archimedes' screw to raise water.*

Ancient Greece

THE ATOM AND NUCLEAR POWER

The ancient Greeks were the first to speculate that the world is composed of atoms. Atoms are the tiny particles, too small to be seen, that make up all matter. One ancient Greek, called Democritus (c.460–370 BC) taught that the world is made up of two things: atoms and void. He said that atoms cannot be cut up (which is what the word 'atom' means) and that they are always moving.

These first speculations were the foundations of modern nuclear physics and chemistry. The 'nucleus' is the centre of the atom and since the beginning of the twentieth century, scientists have been finding out about the enormous amount of energy which is locked there. When the atom is split, a reaction happens which releases nuclear energy. This energy was first

▲ *Greek philosophers, like Democritus, were the first people to ask big questions about the workings of the universe. Before their time, people had always looked to religion to explain how the world was made and and what it was made of. The ancient Greeks were the first true scientists.*

▶ *This is the heart of a modern nuclear power station.*

30

released in a nuclear reactor test, held in a squash court at the University of Chicago, USA, on 22 December 1942. Since then, most developed countries have started to generate electricity with nuclear energy, which is based on the understanding of atoms first discovered by the ancient Greeks.

STEAM ENGINE

Steam is the vapour that water turns into when it boils. Steam under pressure can be used as a power to drive machinery. The first steam engine was designed by Hero of Alexandria in the first century AD. The 'whirling ball', or *aeolipyle*, produced steam power by first heating water in a cauldron. The steam from the water would pass up a pipe into a hollow globe suspended above the cauldron. Two small openings in the globe would allow the steam to escape in opposite directions, which would cause the globe to spin on its axis. Hero also designed a larger device, based on the same principles, to open temple doors.

Steam power is used today to drive generators in nuclear power stations and to power naval and merchant ships. The first nuclear submarine, the *USS Nautilus*, surfaced at the North Pole on the 3 August 1958. It was driven by a steam turbine using heat from a nuclear reactor, which propelled it 330,000 nautical miles in ten years.

Hero's Steam Turbine

▲ As well as being the first steam engine, Hero's device used an early form of jet power. The globe was turned by the backward thrust, or jet, of steam.

chapter six

ANCIENT ROME

The Romans were highly skilled in the organization of their huge empire and many legacies of technology have been left as a result. The administration of the Roman Empire, along with its roads, engineering and architecture, have all influenced the way we do things today.

HOSPITALS

One of the legacies of the Romans is the setting up of a hospital system. The Empire grew by using its huge army to conquer foreign lands. At first, injured soldiers were sent home for treatment. But as the Empire's frontiers became more and more distant, it was too far to send soldiers home for treatment. So military hospitals were set up. One of these hospitals, on the lower Rhine river, near Düsseldorf, Germany, has been carefully excavated to show long corridors with wards, dining areas and administrative offices. The ancient hospital also had an excellent drainage system. It is very similar in design to a modern hospital.

Water pipes
The water supply of ancient Rome used both earthenware and lead pipes. This caused many Roman plumbers to be affected by lead poisoning, which made their skin unusually pale in colour. Our word 'plumber' comes from the Latin for lead, *plumbum*.

SEWERAGE

The Romans had a sophisticated system of sewers for their time. The main drain of Rome, called the Cloaca Maxima, acted as the main sewage channel. It wound through the city, carrying the sewage to an outlet in the River Tiber. In a modern system, sewage passes through many stages, including being filtered and treated with chlorine to make it microbe-free.

WATER SUPPLY

Roman householders had water supplied to their homes through pipes and a tap or spout called a 'calix'. They paid a water rate according to the size of the calix. There were twenty-four different sizes of calix. In AD 96, Frontinus, who had been a military governor of Britain, was put in charge of Rome's water. With a staff of 700 men, 900 million litres were distributed through 400 km of piping, filling public cisterns and supplying private households.

▲ This is the mouth of the Cloaca Maxima, or 'great sewer', on the River Tiber in Rome, Italy. It was built 2,500 years ago.

◀ A modern sewage treatment works, showing its various stages of treatment, photographed from a safe distance.

First vets

The Romans were the first to study animal illnesses, and to find ways to cure sick animals. This is called veterinary medicine. The poet Virgil described animal diseases and treatments in great detail. The writer Flavius Vegetius Renatus (AD 383–450), the father of veterinary science, was the author of the first veterinary book to be printed, 1,000 years after his death.

WATER POWER

Water is used to make power today in hydroelectric power stations, where the energy produced from fast-flowing water is used to make electricity. There are more than sixty hydroelectric power stations around the world today, which generate over 1,000 megawatts of electricity a year. In the USA, water is the fourth-largest source of energy. It is a valuable source of power because unlike oil, it is a renewable form of energy, and there is much more water power available than can be used.

The Romans were using water for power by the fourth century BC. A flour mill, built at Barbagel in France in the fourth century BC, used eight pairs of waterwheels (wheels driven by running water) to generate power for its grinders. The mill's grinders could grind about 10 tonnes of grain a day into flour. The water-powered flour mill was the most powerful machine of the ancient world. It generated about 30 horsepower, which is equivalent to the power of about ten petrol-driven garden mowers. The flour from the mill supplied the entire population of the nearby town of Arles, which included a Roman garrison and numbered about 10,000 people.

▶ This ancient Roman aqueduct, at Hama in Syria, is still used to transport water from the river to fields around the town. The power of the river turns the wheel, which scoops up the water.

▼ *A modern use of water power to make electricity, this modern hydroelectric dam is in the Andes mountains of Venezuela, South America.*

CIVIL ENGINEERING

Many Roman roads, bridges, aqueducts and dams are still in use today, due to the technology and engineering skills used to build them. Their arched, stone bridges lasted so well after the fall of the Roman Empire in AD 476 that for centuries there was no need to construct any new ones. The Romans developed the arch as a building technique that could be used to build very large constructions such as huge halls and bridges. Stone arches were built by first making an arch shape out of wood, which supported the shape as stone was added. When all the stone was in place, the wood could be taken away. Arched bridges, such as the Pont du Gard bridge at Nîmes, in France, were made by first building columns of stone blocks. The blocks were lifted into place using a crane powered by a treadmill. The columns were filled in with rubble or small, rough stones. Arches were then built between each column.

▼ *The Pont du Gard bridge was built by the Romans across the River Gardon in France. It is 50 m high. The top level was an aqueduct, for carrying water. Below is the bridge for travellers.*

ROAD BUILDING

The Romans built 80,000 km of paved roads which were constructed using successive layers of rubble, crushed stone and sand. The roads had very deep foundations, often 1 m thick, and they were 10 m wide. Many modern roads today are still built in layers. The Romans used the roads to transport their army and trade around their Empire.

CONCRETE

The Romans discovered how to make concrete in the third century BC, by mixing together volcanic ash, lime, sand, gravel and water. This mixture dried to become a type of artificial rock. It was waterproof and solid and could be set into any shape. The Romans poured it into brick moulds and used it in the construction of buildings. They used concrete in the foundations, walls and vaults of the vast Colosseum, a famous amphitheatre in Rome, Italy. Modern concrete is still an important material used in building construction today. It is also used to make motorways, where continuous slabs measuring 10 m wide are laid down.

▲ *This modern motorway overpass in Texas, USA, uses concrete, another Roman invention.*

TIMEKEEPING

Today, the quickest way to tell the time is to look at a clock or wristwatch. The ancestor of modern mechanical clocks is the sundial, first invented by the ancient Egyptians. The Romans invented portable sundials to be used by travellers. Roman sundials consisted of an engraved metal disc, fitted with a curved, raised arm which cast a shadow. Certain times of the year were marked on the disc. The Romans also invented a water clock to tell the time. The clock worked using water that dripped at a constant rate into a reservoir of water. A rod, floating in the water, moved upwards as the water level increased, and moved the hand of a clock through a cog.

FIRST MILOMETER

A milometer is a device that records the number of miles that a bicycle or motor vehicle has travelled. The first milometer, called a 'hodometer', was invented by the Romans about 100 BC, and was written about by the Roman architect Vitruvius. The hodometer consisted of a wheel that ran along the ground, turning a series of cog wheels. A disc with holes in it carried pebbles within the hodometer, and for every mile that was travelled, one pebble fell into a reservoir at the bottom of the device. The number of miles travelled could either be measured by counting the number of pebbles in the reservoir, or by looking at the dials on the device. Modern milometers work electronically by recording the number of times a vehicle's wheels have turned.

CENTRAL HEATING

When the Roman Empire expanded north to Britain, France and Germany, Romans found that their houses had to be heated so that people could keep warm. So they invented the hypercaust, a system of underfloor heating. The heat and fumes from a fireplace at one side of a room or building passed through stone channels that ran under the floor. These channels, or flues, then led to the chimney. Cavity walls (double walls with an air space in between) were also constructed, and sometimes the warm air passed through these as well. The same system could be used to heat the Roman baths. The hypercaust is the ancestor of modern central heating.

▼ You can still see the Roman underfloor heating system at the palace of Fishbourne in England.

◄ Solar power provides the newest way to solve the ancient problem of heating our homes. The panels on these roofs in Saro, Sweden, collect and store the sun's heat. The advantage of this system is that it causes no pollution, and it does not use resources such as coal and oil, which cannot be replaced.

Ancient Americas

chapter seven

ANCIENT AMERICAS

▼ Little is known about the ancient Olmecs, who carved these huge statues using only stone tools. We do not even know what they called themselves. The name 'Olmec' was given to them by an American museum director in 1929.

MONUMENTAL SCULPTURE

Mexico's first great civilization, that of the Olmecs, appeared on the coast of the Gulf of Mexico between 1500 BC and 1200 BC. Unlike other ancient civilizations at the same time, the Olmecs did not have the help of the wheel to pull carts, or iron tools to build with. Despite this, they managed to produce monumental sculptures, huge religious centres and complicated irrigation systems. They carved colossal heads out of single basalt blocks, some weighing as much as 20 tonnes. These monuments were carved as portraits of Olmec rulers, and they still surround the ancient centres of La Venta and San Lorenzo, in Mexico. The basalt was taken from the Tuxtlas Mountains, 90 km away. This ancient tradition of carving giant statues of rulers has been continued by the modern American sculptor, Gutzon Borglum. His gigantic heads of presidents Washington, Jefferson, Lincoln and Roosevelt can be seen at Mount Rushmore in South Dakota, USA.

RUBBER

The Olmecs were probably the first to make use of rubber for clothing. They had a national ball game that was so fast that the players had to wear protective clothing. The game was played with a large rubber ball. The rubber came from the Havea tree, and was used to make both the ball and the protective clothing.

▼ These huge statues of four famous American presidents are carved on to the mountainside of Mount Rushmore, in North Dakota, USA.

> **First grown**
> Tomatoes, avocadoes, beans and pumpkins all first came from ancient Central America. Potatoes and manioc are the staple crops of the Amerindian people of South America. Tobacco, cocoa and quinine are valuable products, now grown world-wide, which also came from this region.

RELIGIOUS CENTRES

Great pyramid-temples were built in ancient Central and South America for religious ceremonies. Religious centres have been designed on a grand scale right up to modern times, for example huge cathedrals, abbeys and mosques. Some of the largest pyramid-temples in ancient Central and South America were built by the Mayan people of ancient Mexico, between 100 BC–AD 800. Mayan temple-pyramids were built of stone, and they often had steps or terraces going up the outside. The pyramid usually had a flattened top, on which a temple chamber was built. The Mayan city of Tikal, in Guatemala, had five temple complexes, each with a pyramid of up to 70 m high. The Pyramid of the Giant Jaguar had a small chamber at the top, where priests performed religious ceremonies. The shape of the pyramid meant that the voices of the priests were carried down so that they could be heard at the bottom of the pyramid.

MAIZE CULTIVATION

The Moche people of northern Peru (about AD 50–700) developed irrigation schemes which

▼ The great Mayan Pyramid of Chichen Itza, with its temple building on the top. It was dedicated to a god called Kukulcan, the feathered serpent.

◀ People have always built religious buildings that soar upwards. This modern church in Iceland looks like a space rocket.

allowed them to cultivate maize, sweet potatoes, peanuts and peppers. Just as wheat was specially cultivated in the Middle East, the careful cultivation of maize, which was already grown by 5000 BC, produced a plant with a far heavier crop than the wild varieties. So the next time you splash cold milk on to your breakfast cereal, just think that the cereal crop you are eating – maize – has been eaten continuously for more than 7,000 years!

Time Line

The Middle East and Egypt

Before 3000 BC	2000 BC	1000 BC	0	AD
7000 First farmers.	**2800** Bronze alloys used in Mesopotamia.	**2000** Shadow clocks used by Egyptians.	**300** Alexandria is a centre of scientific learning.	**100** Steam machine designed by Hero of Alexandria, Egypt.
3800 The wheel is invented in Sumer and used to make pottery and to move carts and chariots.	First calender of 365 days invented by the Egyptians.	**1520** Glass bottles appear in Egypt.		
3600 Kilns used in Sumeria for pottery.	**2400** Simple looms used in the Near East.	**1440** An Egyptian water-clock shows the hours.		
3400 Primitive wooden plough used in Egypt and Sumeria.	**2400** Sewage and drainage systems in the Indus Valley (modern-day India and Pakistan).	**1360** Egyptians chariots have spoked wheels.		
3200 Bronze axes moulded in Sumeria.		**1100** Iron plough used in Mesopotamia.		

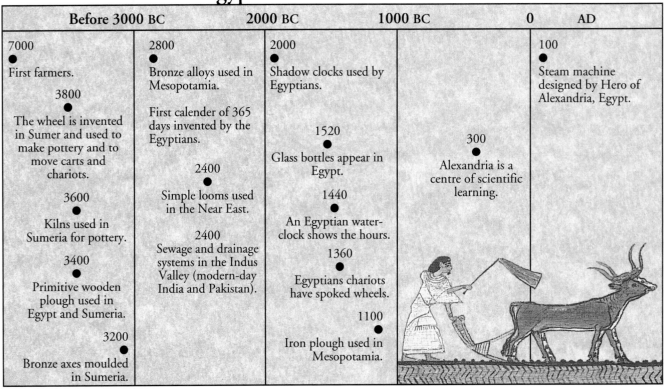

The Far East

Before 3000 BC	2000 BC	1000 BC	0	AD
9500 First pottery, Japan.	**2700** Silkworms being used to make silk in China.	**1900** First cities built in China.		**100** Chinese astronomers record eclipses.
5000 Rice grown in China.	**2500** Bronzeworking in Southeast Asia.			**190** Abacus used in China.
3000 Potter's wheel developed in China.				**200** First seismograph invented in China by Chang Heng.
				430 Many Chinese scientific instruments developed.

Time Line

Europe and the Mediterranean Lands

Before 3000 BC	2000 BC	1000 BC	0	AD
6000 — Copper made into beads in northern Europe. 3700 — Bronze Age starts in parts of northern Europe. 3000 — Bronze Age culture begins in Greece.		1600 — Building of Stonehenge in England is completed. 1200 — Greeks and Phoenicians preserve fish using a drying and smoking technique.	470–400 — Life of the Greek doctor Hippocrates. 460 — Democritus discovers atoms. 287–212 — Life of Archimedes of Syracuse, Sicily. 230 — Eratosthenes measures the size of the earth. 60 — Roman watermills in operation.	20 — Roller bearings for wheels are developed in Denmark. 100 — Doctors in Rome use surgical instruments. 160 — Galen writes a detailed medical book.

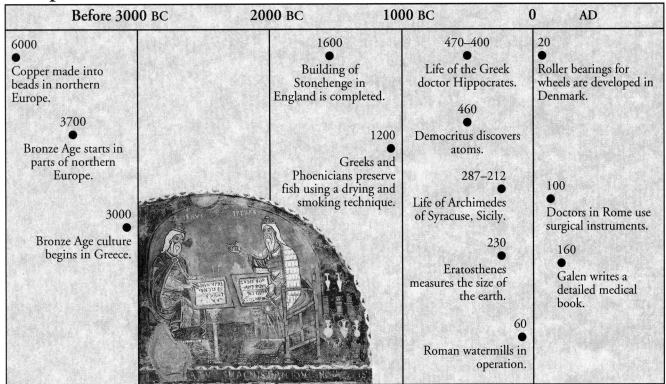

The Americas

Before 3000 BC	2000 BC	1000 BC	0	AD
3000 — First-known pottery in Ecuador.	2300 — Pottery in use in Mexico and Guatemala. 2000 — Manos Cruzados temple built in Peru.	1500 — Olmec civilization begins in Mexico. 1200 — Olmec capital founded at San Lorenzo, Mexico.	900 — New Olmec capital founded at La Venta, Mexico. 750 — First goldworking in South America. 600 — First ball-courts built at the Olmec centres of San Lorenzo and La Venta. 250 — Rise of Maya civilization in Central America.	340 — Mayan calendars and astronomy develop.

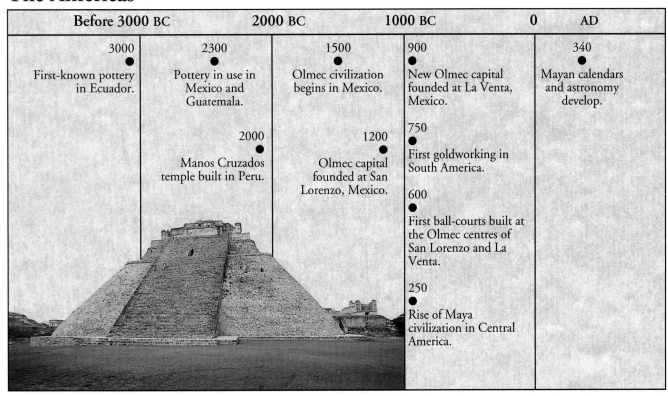

GLOSSARY

Axis A real or imaginary line through the middle of an object around which it rotates.
Axle The rod or pin on which a wheel turns.
Cisterns Tanks for storing water or other liquids.
Corroded Worn or eaten away.
Crop Grain, fruit or vegetables once they are grown.
Cross-breeding Reproduction from two different types of plant or animal.
Density A measurement of how thick, or compact an object is.
Displace To cause a quantity of liquid to move.
Domesticated Tamed.
Draught animals Animals that are used to pull loads.
Friction Rubbing of the surface of one object against another.
Horsepower Measurement of power: 1 horsepower = 754 watts of electricity – this is actually more power than a single horse can produce.
Kilowatts (kW) Measurement of electric power: 1 kW = 1,000 watts.
Megawatts Measurement of electric power: 1 megawatt = 1 million watts.
Micro-organisms Living animals or plants that are so small they can only be seen with a microscope.

Microwave oven A high-tech machine that cooks food using waves of radiation.
Ores Rocks or minerals from which metals can be obtained.
Pharaohs Kings in ancient Egypt.
Physician Doctor.
Quinine A bitter medicine, made from the bark of a South American tree, used to treat malaria and other fevers.
Roller bearings Balls within a wheel that bear the friction of the part that is moving.
Sluices Artificial channels for water, with a sliding gate to control or change the direction.
Spokes Rods or bars that are fixed between the centre of a wheel and its rim.
Synthetic Artificial.
Treadmill A large wheel turned by the weight of people or animals treading on steps fixed all round it.
Turbine An engine powered by steam or gas which forces wheels with blades attached to revolve continuously.
Volume The amount of space that is taken up by an object.

BOOKS TO READ

Famous Inventors by Douglas McTavish (Wayland, 1993)
History Makers of the Scientific Revolution by Nina Morgan (Wayland, 1995)
Invention (Eyewitness Guides) by Lionel Bender (Dorling Kindersley, 1991)
Inventions in Science: The Car by Steve Parker (Watts, 1994)
Pioneers of Science: Archimedes by Peter Lafferty (Wayland, 1991)
The Ancient World by John Briquebec (Kingfisher, 1990)
The Atlas of Ancient Worlds by Ann Millard (Dorling Kindersley, 1994)
The Atlas of the Ancient World by Margaret Oliphant (Ebury Press, 1992)
The Builder through History by Richard Wood (Wayland, 1994)
The Earliest Civilizations by Margaret Oliphant (Simon & Schuster, 1993)
The Hamlyn All-Colour Science Encyclopedia (Hamlyn, 1994)
The Inventor through History by Peter Lafferty and Julian Rowe (Wayland, 1993)

Picture acknowledgements:
The publishers would like to thank the following for allowing their pictures to be used in this book:
AKG London 4 (middle right), 24, 45 (top); Art Directors Photo Library 5 (top), 23; Bridgeman Art Library (Index, Barcelona) 16; Delta Archive *Cover* (main), 13 (main); C M Dixon *Cover* (inset), title page (inset), 5 (bottom), 7 (bottom), 12, 14-15 (bottom), 18, 33 (top), 39 (top), 44 (top); Eye Ubiquitous 4 (top), 19, 28-9 (bottom), 32-3 (bottom), 36; Robert Harding 4 (bottom), 8 (top), 15 (top), 22, 34, 44 (bottom); Michael Holford 21 (top); Images Colour Library 41; Science Photo Library 10-11 (bottom), 28 (top), 30 (top), 39 (bottom); Square Studio 13 (top); Tony Stone Worldwide 6, 7 (top), 8-9 (bottom), 14 (top), 17 (top), 21 (bottom), 25, 27, 30 (bottom), 35, 37, 40, 42-3 (bottom); Wayland Picture Library 4 (left), 45 (bottom); Werner Forman Archive 10 (top).
All artwork is by Peter Bull.

INDEX

Numbers in **bold** refer to illustrations.

abacus 22, **22**, 44
Amerindian people 42
Archimedes 28–9, **28**, 45
  screw 28, **29**
architecture 16-17, **17**, 32
astronomy 13, 19, 26–7, **27**, 45
atoms 30–1, 45

Babylonians 13
beer 18
bread 18, **18**, 19
bridges **4**, 36, 37

calculations 22, **22**
calendars 13, 19, 44, 45
car **1**, 13, **13**
chariots **1**, **5**, 12, **12**, 44
civil engineering 17, 36–7
clocks 38, 44
clothing 6, 7, 41
Colosseum 37
compass 20, **20**
concrete 37
construction 16–17, **16**, **17**

dams **10**, **10–11**, 11, **35**, 36
Democritus 30, **30**, 45

Eratosthenes 26

farming **4**, 8, 8–9, **8–9**, 19
fire 6–7, **6**, **7**
fireworks **5**, 22, 23
fishing 10
France **4**, 34, 36

Galen **4**, 24, 25
genetic engineering 9
Germany 32
Guatemala 42

heating 6, **38–9**, 39
  hypercaust 39, **39**
Heng, Chang 21, 44
Hero of Alexandria 31, 44
Hiero, King of Syracuse 28, 29
Hippocrates **4**, **24**, 24–5,
hospitals 32
hydroelectricity 34, **35**

Iceland 43
irrigation 10–11, 42-3

Japan 22, 44

Maya **4**, 42, **42–3**
medicine 4, 7, 24, **25**, 45
Mesopotamia **5**, 8, 10, 12, 44
metallurgy 7
Mexico **4**, **40–43**
microwave 7
milometer 38
Moche people, Peru 42–3

New Zealand **10–11**
nuclear power 7, 30–31, **30**

Olmecs 40, 41, 45
Pantheon 37
Peru 42, 45
plough 14–15, **14–15**, 44
plumbing 32, 33
pyramids **4**, 16–17, **16**, 42,
  42–3
  Cheops (Great) 16, **16**
  Mayan 42, **42-3**
Pythagoras 27, 45

religious buildings 42–3, **42–3**
rivers
  Columbia 11
  Euphrates 10, 11
  Nile 15, 19

Tiber 33
Tigris 10, 11
roads 32, 36, 37, **37**, 45
roller bearings 12, 45, 46
Rollerblading 13, **13**
Roman
  baths 39
  engineering 36
  road-building 37, 45
Rome 32, **33**, 37
rubber 41

sculpture 40, **40**, **41**
seismograph 20, 21, **21**, 44
sewerage systems 32–3, **32–3**, 44
speedometer 38, 39
steam power 31, **31**, 44
Sumerian civilization 10, 11, 12
sundials 38
Sweden **39**

timekeeping 38

USA 11, 40, **40**

veterinary medicine 34

water power 34, **34**, **35**
  clocks 38, 44
  hydroelectricity 34, **35**
  wheels 34, **34**
wheel 12–13, **13**, 44